TANGRAMABLES
A TANGRAM ACTIVITY BOOK

LEARNING RESOURCES

Written by Judi Martschinke
Drawings by Stephen F. Smith
& David W. Pauli

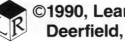
©1990, Learning Resources
Deerfield, Illinois 60015

TANGRAMABLES
Notes to the Teacher

The **TANGRAMABLES book contains three sections**

> **COVER UP**
> **IMAGINATION**
> **SHAPES FROM SHAPES**

COVER UP activities are appropriate for K-1 students but can be completed by students at all grade levels. **IMAGINATION** activities are targeted for students in grades 2-3-4, but again can be completed by all grade levels. **SHAPES FROM SHAPES** activities are more challenging and targeted for grades 3-4-5, although students in grades K-2 can complete many of the tasks. Older students (grades 6-8) will also find some of the pages in the **IMAGINATION** and **SHAPES FROM SHAPES** sections both challenging and enjoyable.

Tangram pieces must be used in conjunction with the **TANGRAMABLES** book. These pieces may be purchased or made. Pieces should be the same size as those pictured on the Tangram Pattern page so students will be able to place pieces directly on the pages in the book when completing the task described.

COVER UP

This section of the book involves placing tangram pieces on top of the outlined tangram pieces. Students need to slide, turn and flip the pieces (transformational geometry concepts) in order to place them correctly. The first few pages involve using only one piece at a time. The pages that follow start combining shapes until, eventually, all seven pieces are used at the same time.

Young children need only to cover the shapes drawn in the book to complete the task. As students get older they should be able to answer such questions as

How many pieces are used?
How many triangles are used?
    squares?
    parallelograms?
How many different size triangles are used?
Which pieces have sides the same length?
Can I fit the piece on top of the drawing if I turn it a half turn?
    a quarter turn?
Can I fit the piece on top of the drawing if I flip it?

IMAGINATION

This section of the book involves combining tangram pieces in such a way that the student's imagination makes the completed work look like something familiar to the student. Drawings of objects or animals are on the pages to help the child use his/her imagination when visualizing the completed work. The outline of an object or animal made by tangram pieces is shown on the page. The child is to cover the area inside the outline with tangram pieces and form the object or animal specified. The outlines of the tangram pieces are shown on the first few pages only. After that the child must use what he/she knows about the pieces (length of sides, size of angles, etc.) and the ways they can be manipulated (slides, turns, flips) to fill in the outline of the object or animal.

Children can use blank paper to lay the tangram pieces on, rather than the **TANGRAMABLES** book. Then they can trace their work for handing in or for displaying. If desired, details can be drawn into the "tangrammable" drawing and background art work can be added.

After using their imagination and experiencing ways to combine tangram pieces in completing "tangrammable" objects, students can make their own "tangrammable" sheets for classmates to try.

SHAPES FROM SHAPES

This section involves making geometric shapes from tangram pieces. Students are asked to use specific pieces or a specific number of pieces to cover the inside of a geometric shape outlined on the page. By doing this students will experience ways various shapes can be formed. Relationships of sides, angles and areas will be experienced when completing the pages. Although students are not asked what they are doing or why, manipulating the pieces will help them understand concepts of length, angle measure, perimeter and area. When the teacher feels students need to measure lengths, measure angles, calculate perimeter, or calculate area, various pages in this section can be completed on blank paper, traced and measured.

Section One
Cover Up

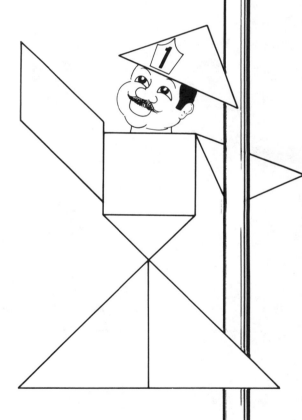

Cover these with the Tangram piece that matches.

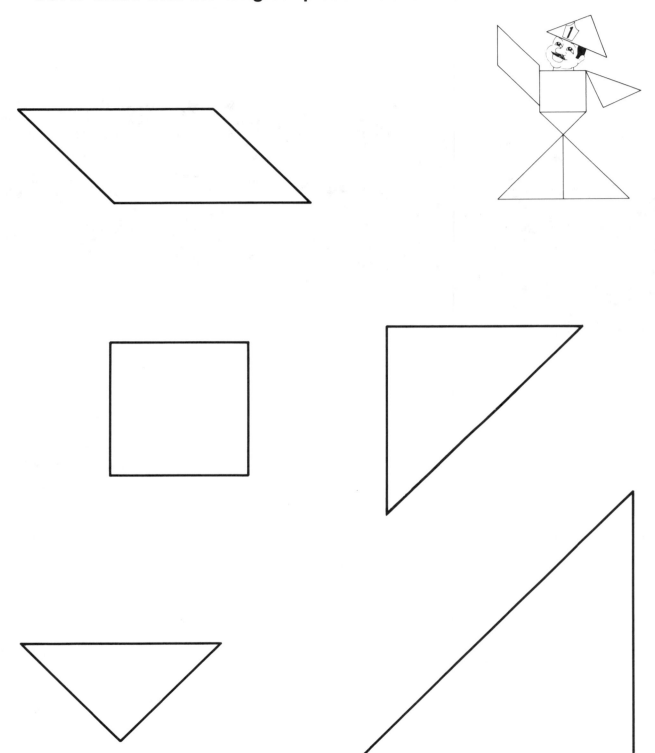

Cover the square with the tangram piece that matches.

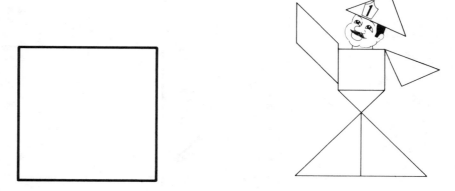

Flip, slide or turn the piece to cover these shapes:

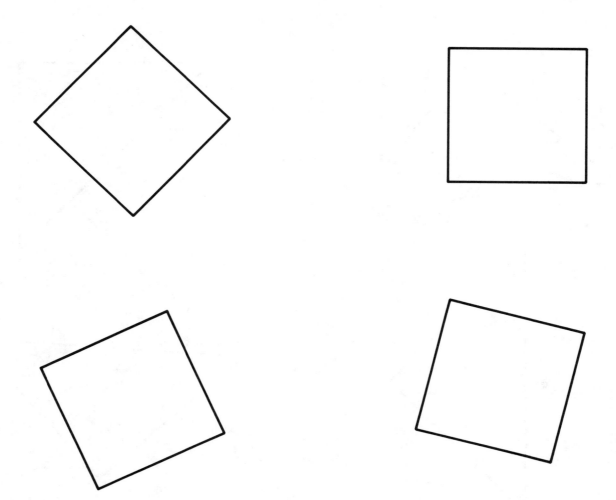

Cover the triangle with the Tangram piece that matches.

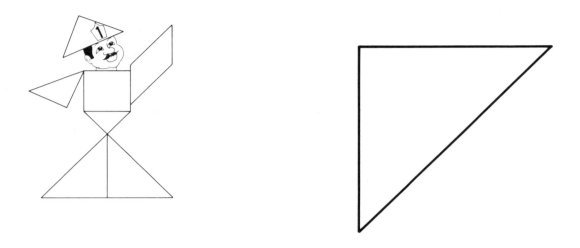

Flip, slide or turn the piece to cover these shapes:

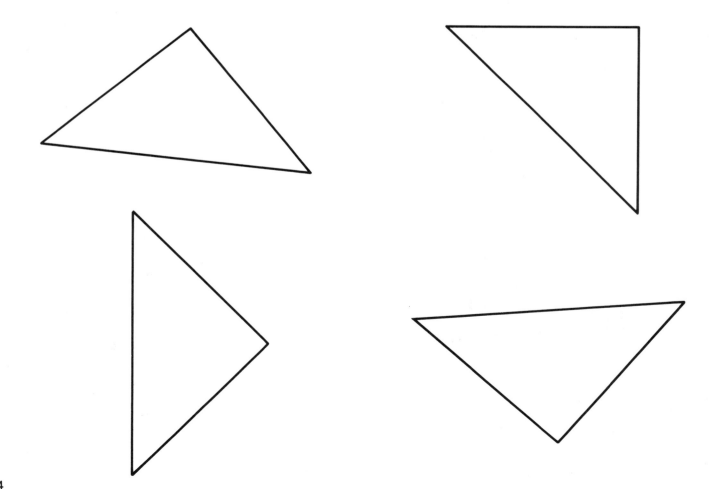

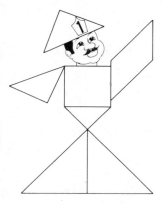

Cover this with the Tangram piece that matches.

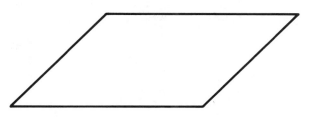

Flip, slide or turn the piece to cover these shapes:

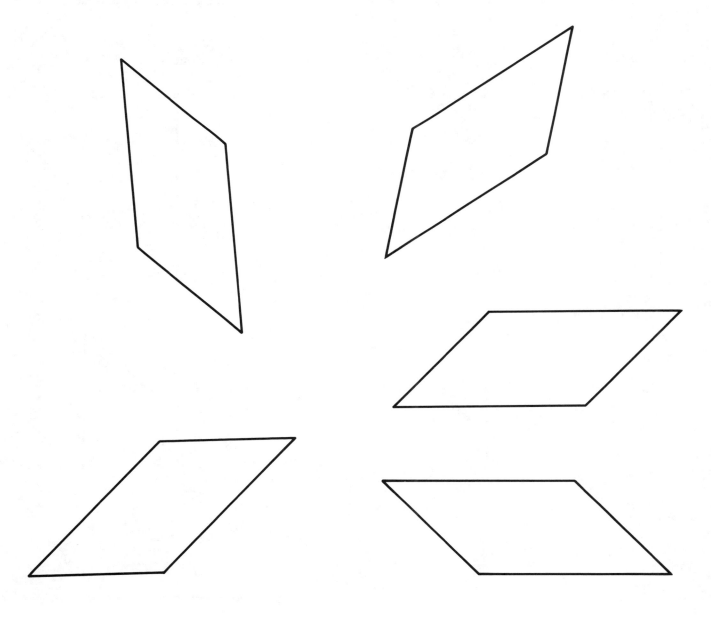

Cover these with Tangram pieces:

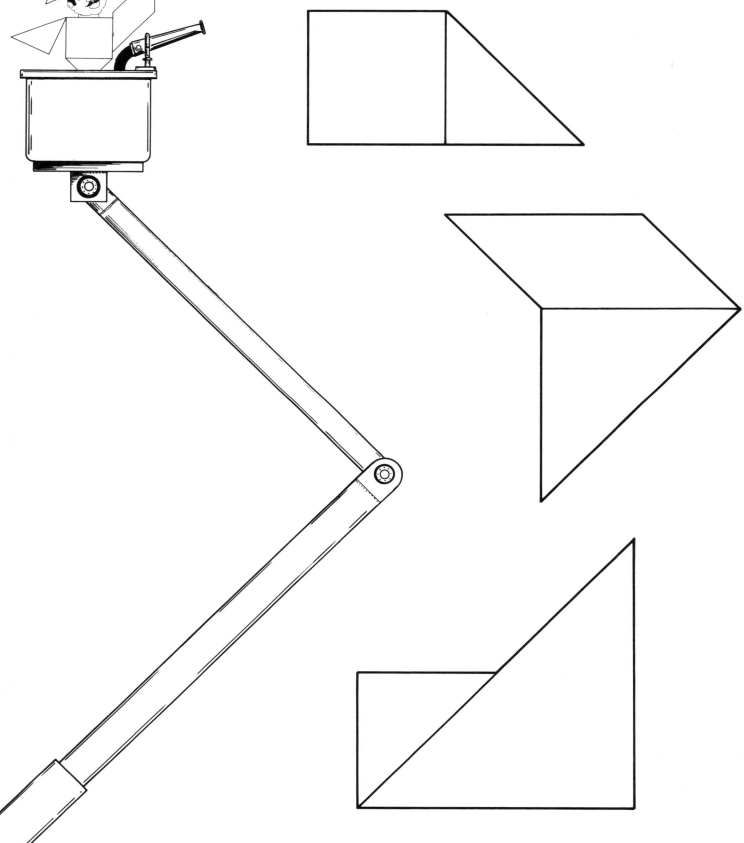

Cover these with Tangram pieces. You will need to use some of the pieces more than one time.

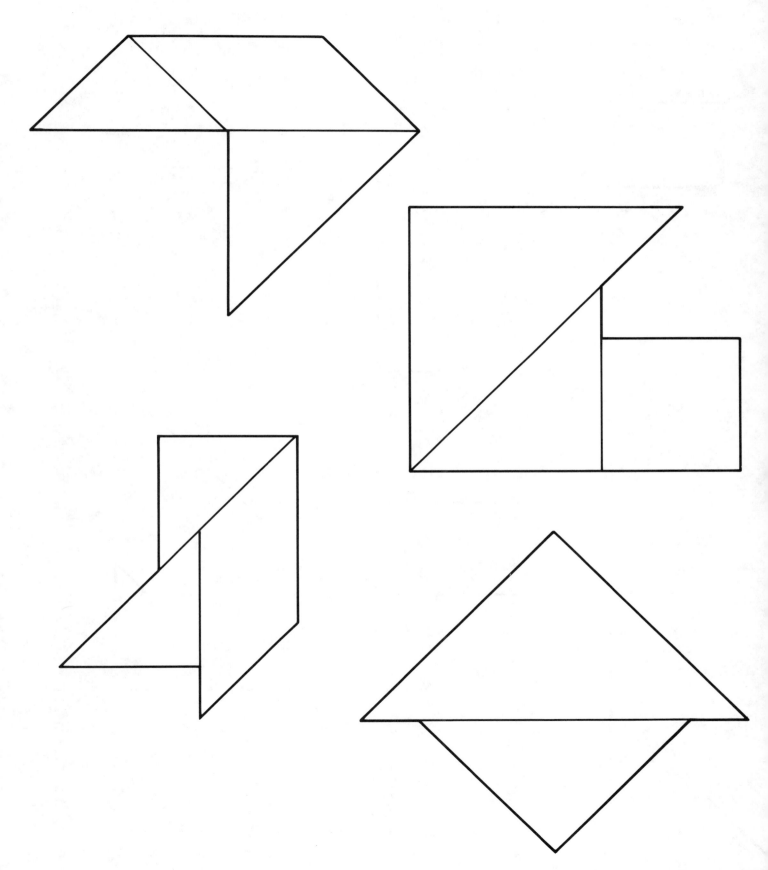

Cover these with Tangram pieces:

Cover this using Tangram pieces:

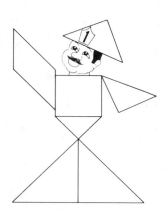

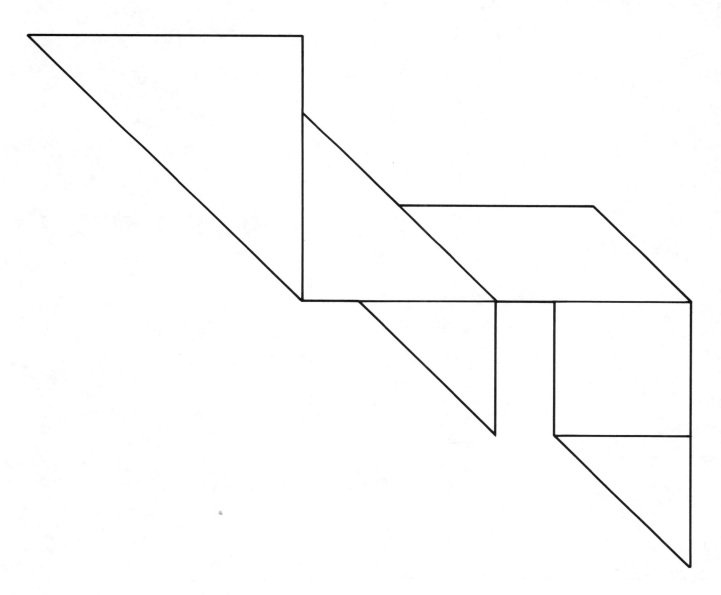

Cover this with Tangram pieces:

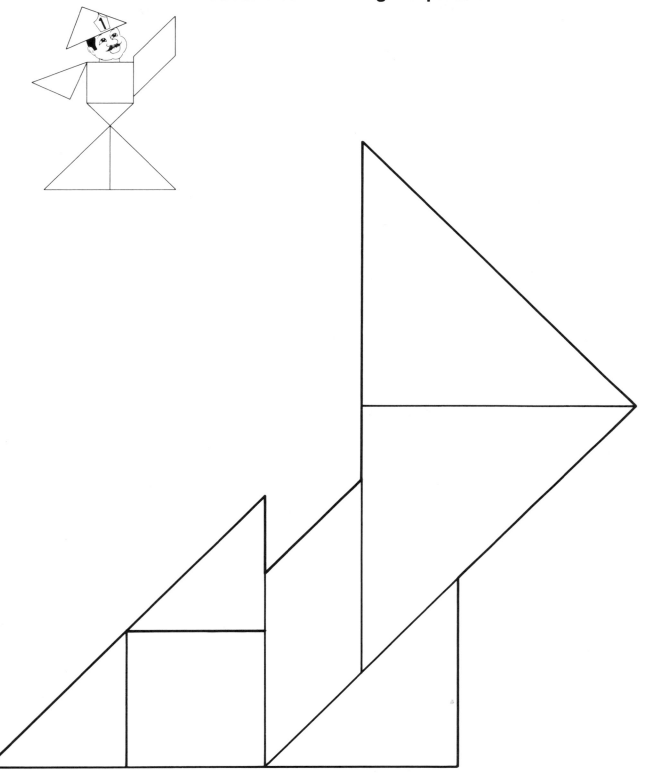

Use Tangram pieces to cover this:

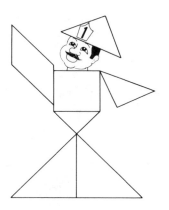

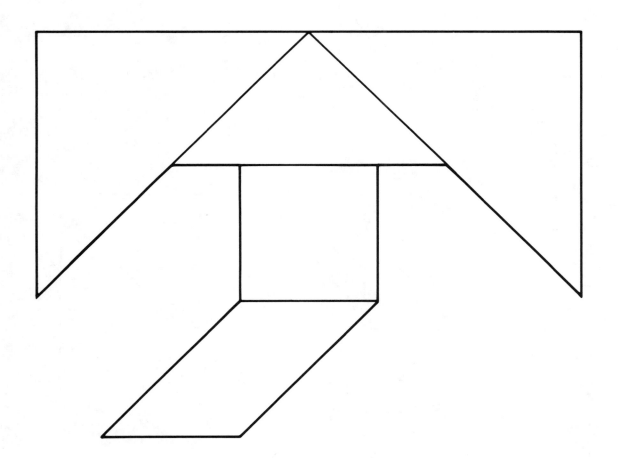

Use Tangram pieces to cover this:

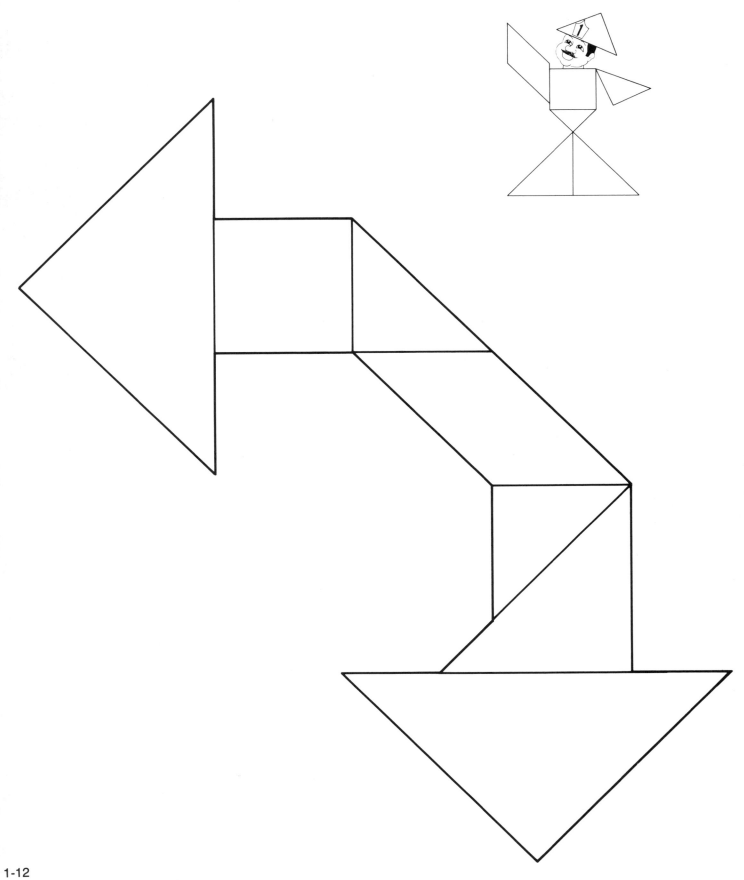

Section Two
Imagination

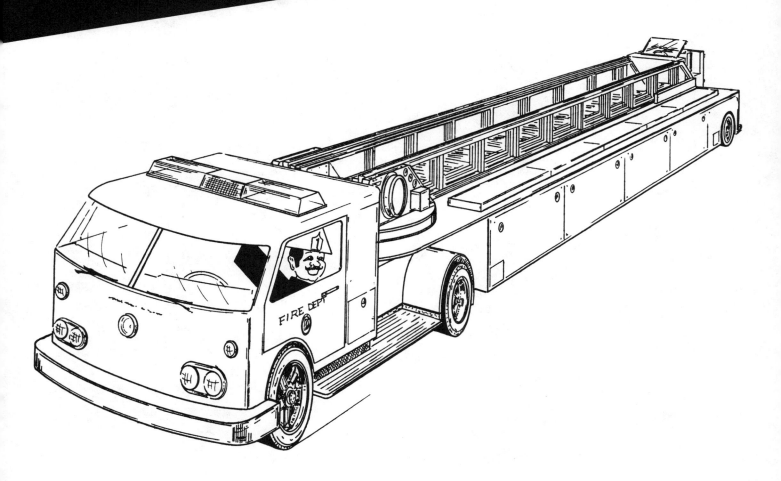

Use your imagination and what you have learned about the way tangram pieces fit together to cover the shapes on the next set of pages.

Use these pieces:

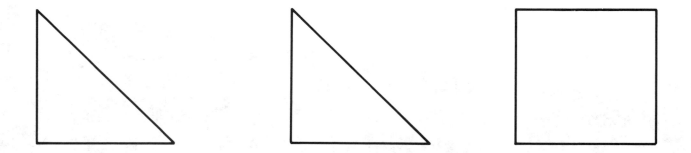

To cover these:

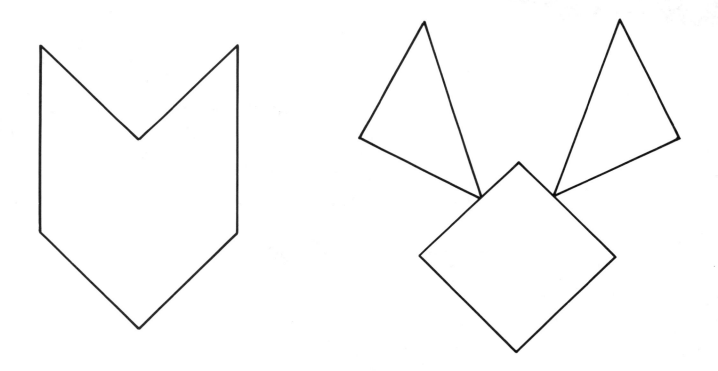

Trace them on your own paper. Draw faces on
them so they look like the heads of animals.

Use 2 pieces to cover this:

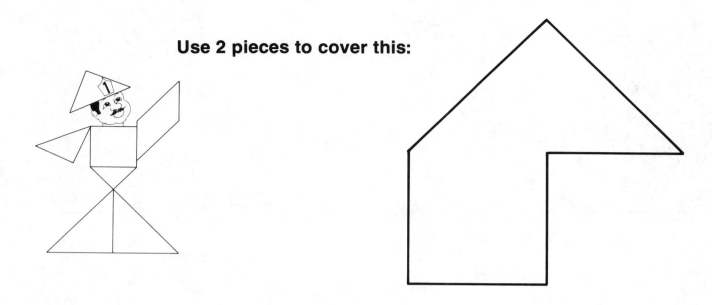

Use 3 pieces to cover this:

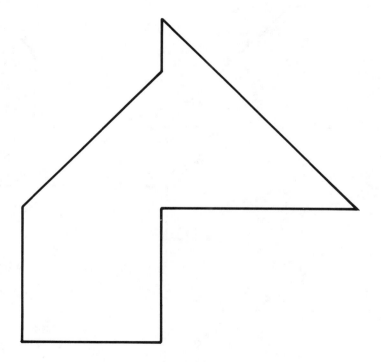

Trace these on your own paper. Draw faces on them so they look like the heads of animals.

Can you make this Rooster using the Tangram pieces?

Make this
"Tangram Kitten."

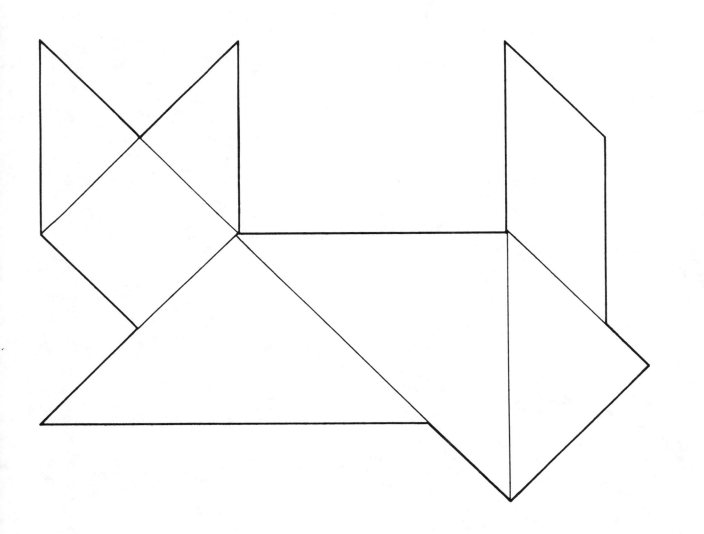

Make this "Tangram Swan."

**Can you make this Cat
using the Tangram pieces?**

**You need all 7 Tangram pieces
to make this Goose.**

Use the Tangram pieces to make this Truck.

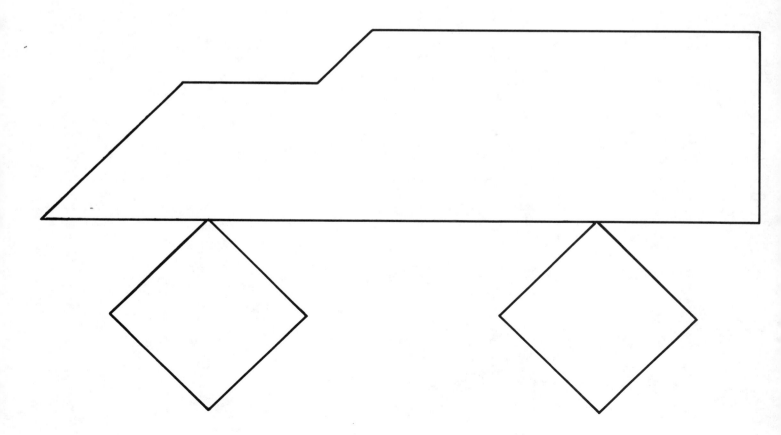

Make this Goldfish using the Tangram pieces.

Can you make this Butterfly using the Tangram pieces?

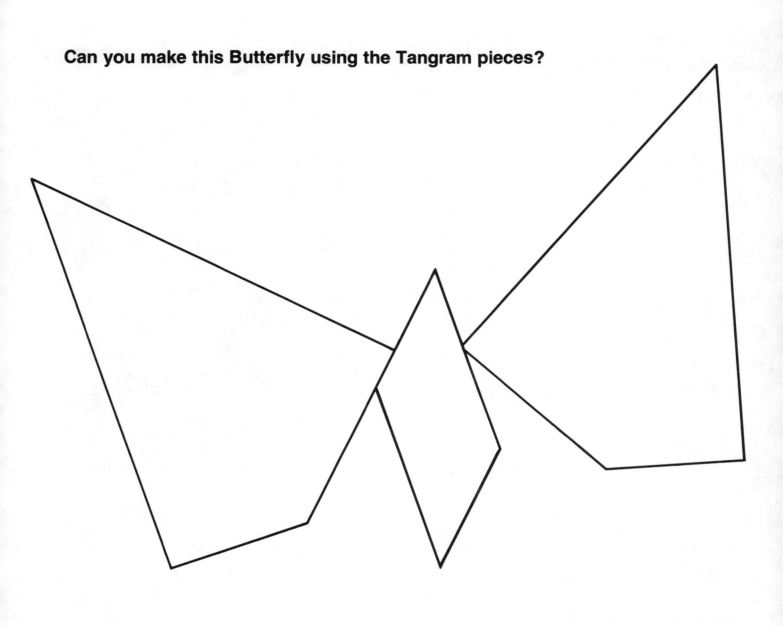

Use the Tangram pieces to make this Lion.

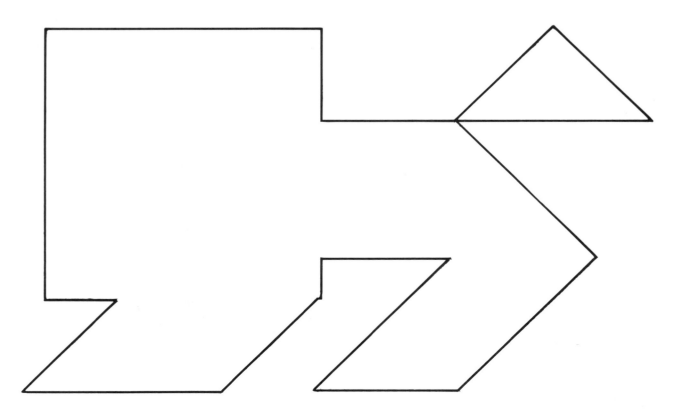

Use the Tangram pieces to make this Teapot.

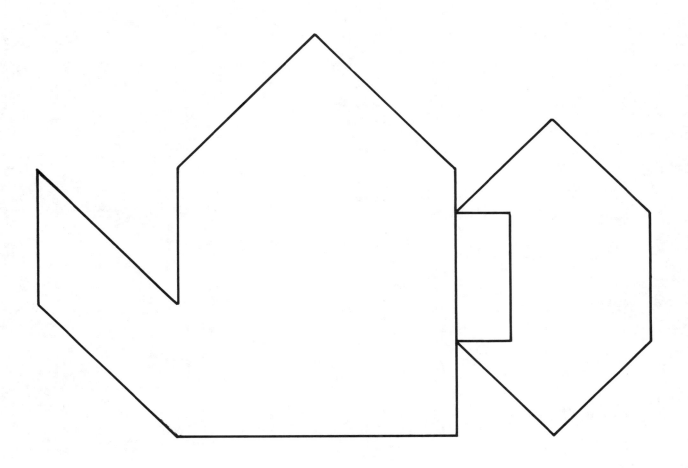

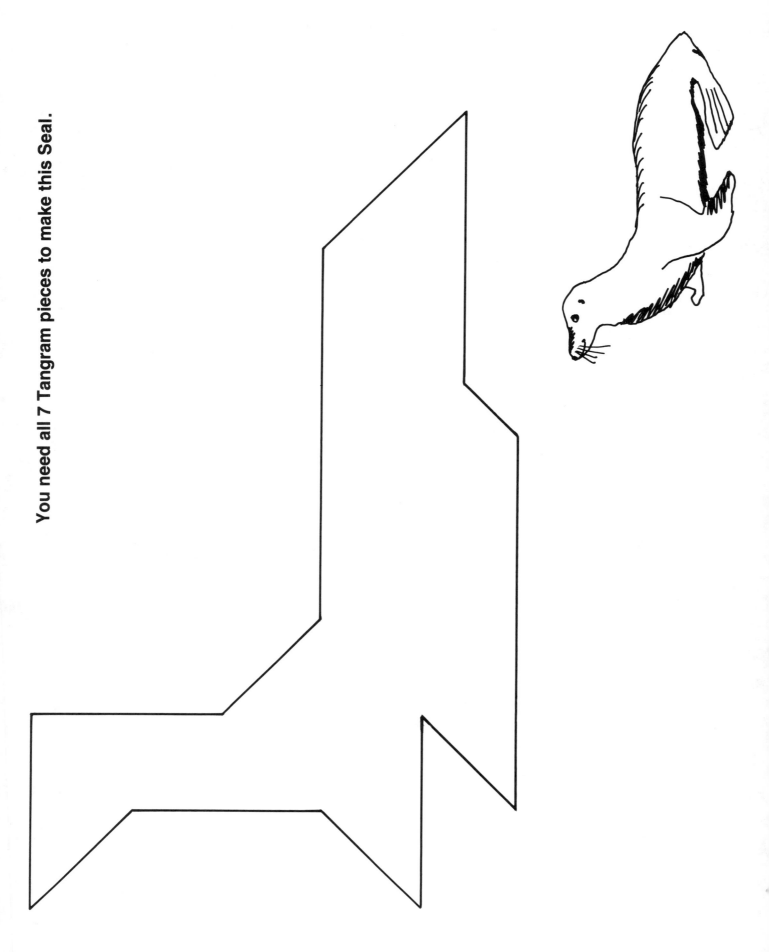

You need all 7 Tangram pieces to make this Seal.

Can you make this Zebra using the Tangram pieces?

You need all 7 Tangram pieces to make this Duck.

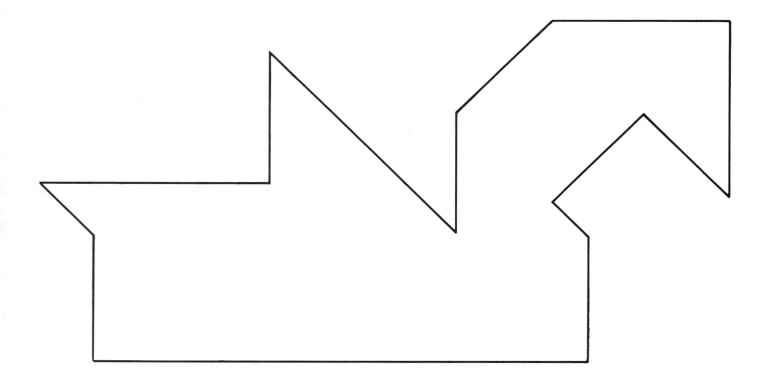

**Make a Camel
like this using
the Tangram pieces.**

Make this Sailboat using the Tangram pieces.

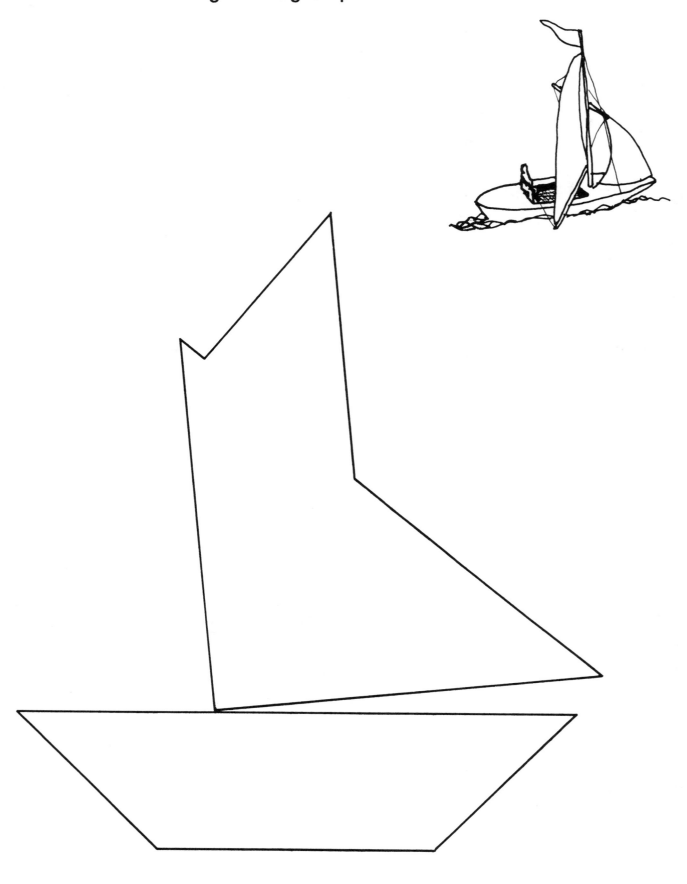

Make a Freighter like this using the Tangram pieces.

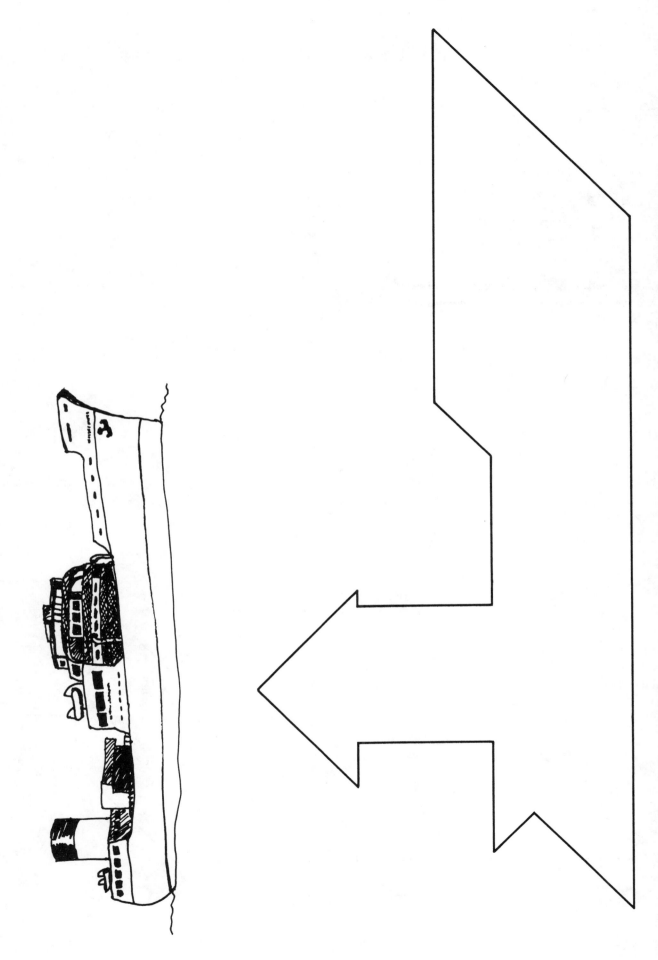

Can you make this Snail using the Tangram pieces?

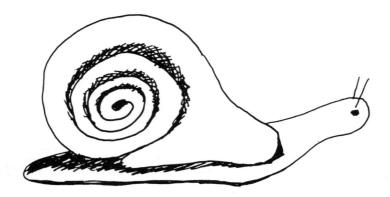

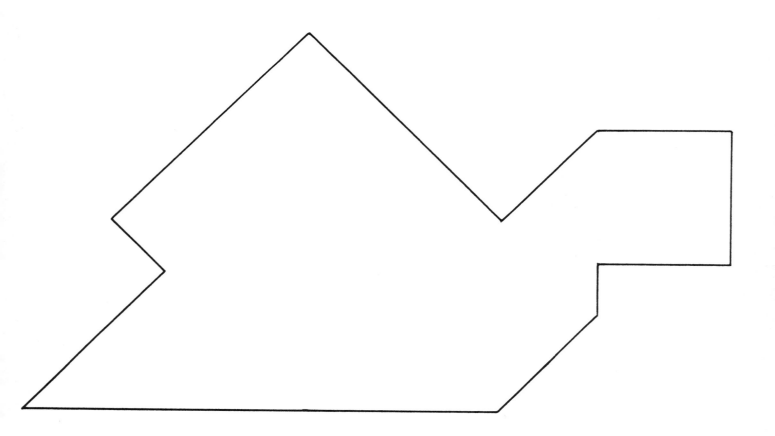

Can you make this Rabbit using the Tangram pieces?

You need all 7 Tangram pieces to make this Goat.

Make this "Tangram Giraffe."

Make a Tugboat like this using the Tangram pieces.

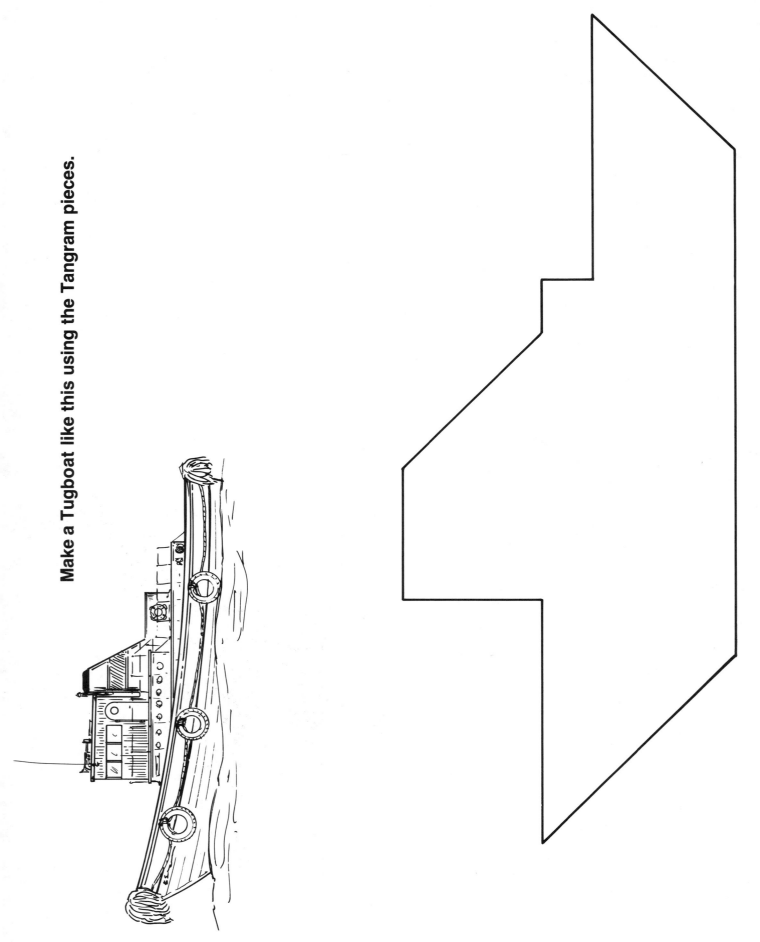

Make this "Tangram Watering Can."

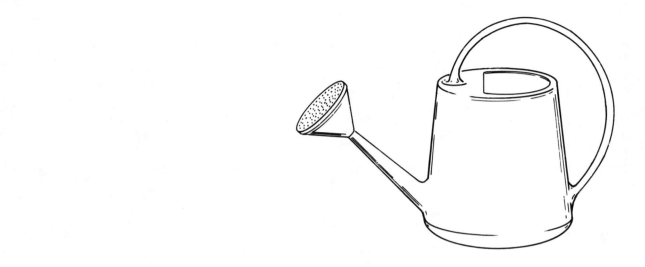

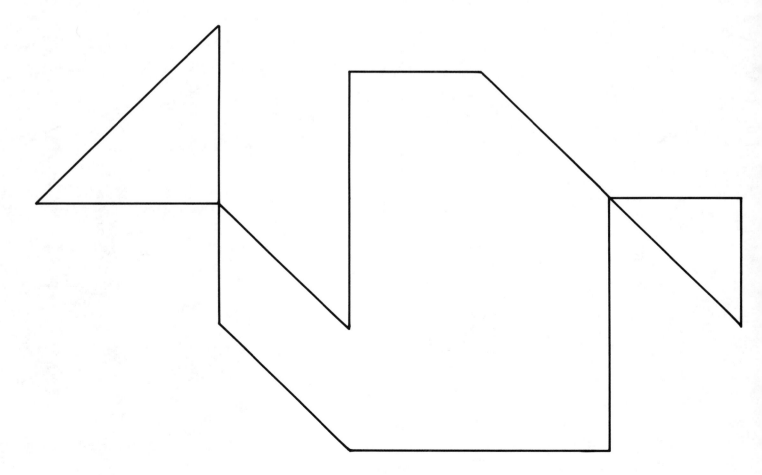

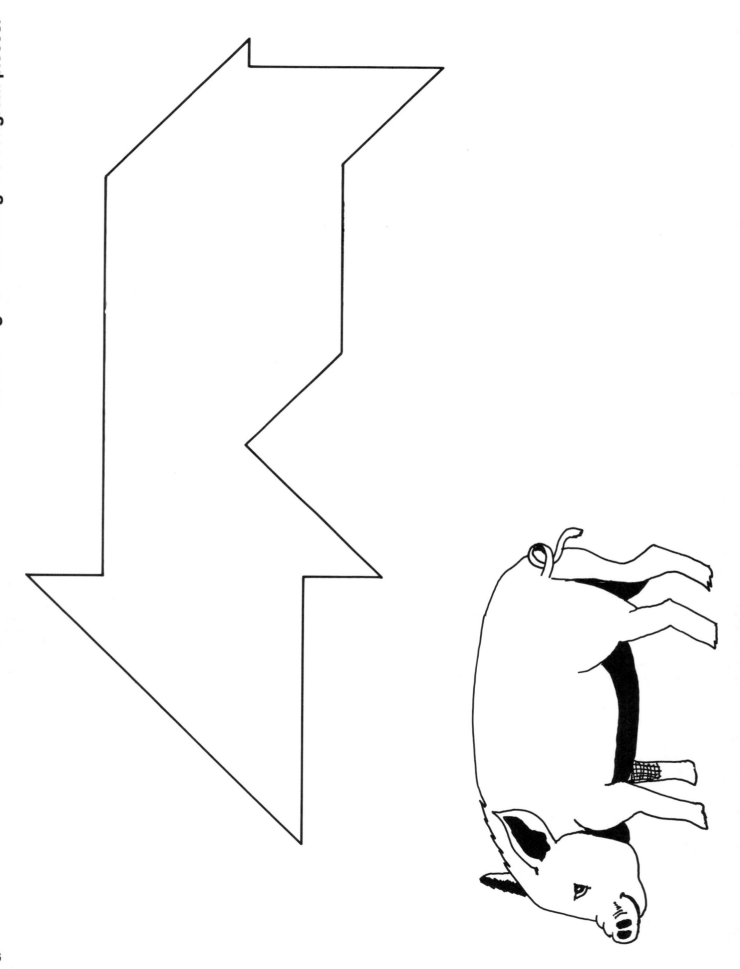

Can you make this Pelican using the Tangram pieces?

**Make this Stork
using the Tangram pieces.**

Make this "Tangram Horse."

Make a Flower Pot like this using the Tangram pieces.

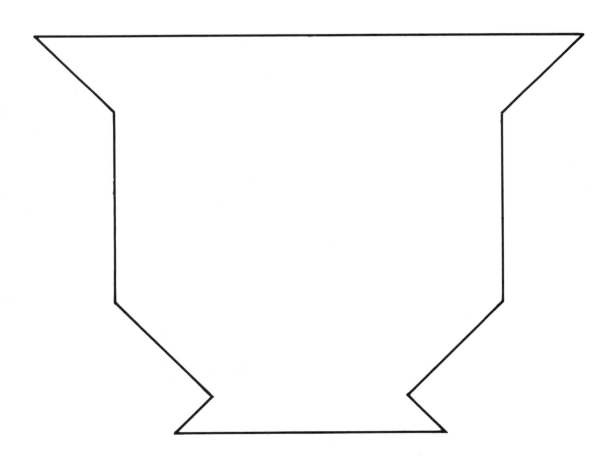

Use the Tangram pieces to make this Jet.

Make this Dog using the Tangram pieces.

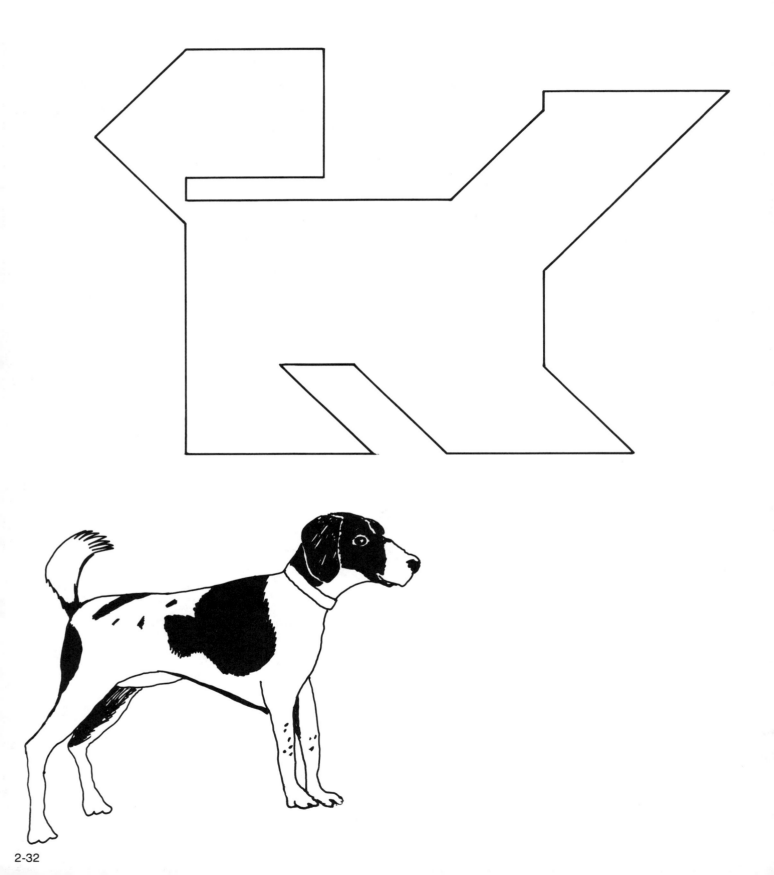

Make this "Tangram Pup."

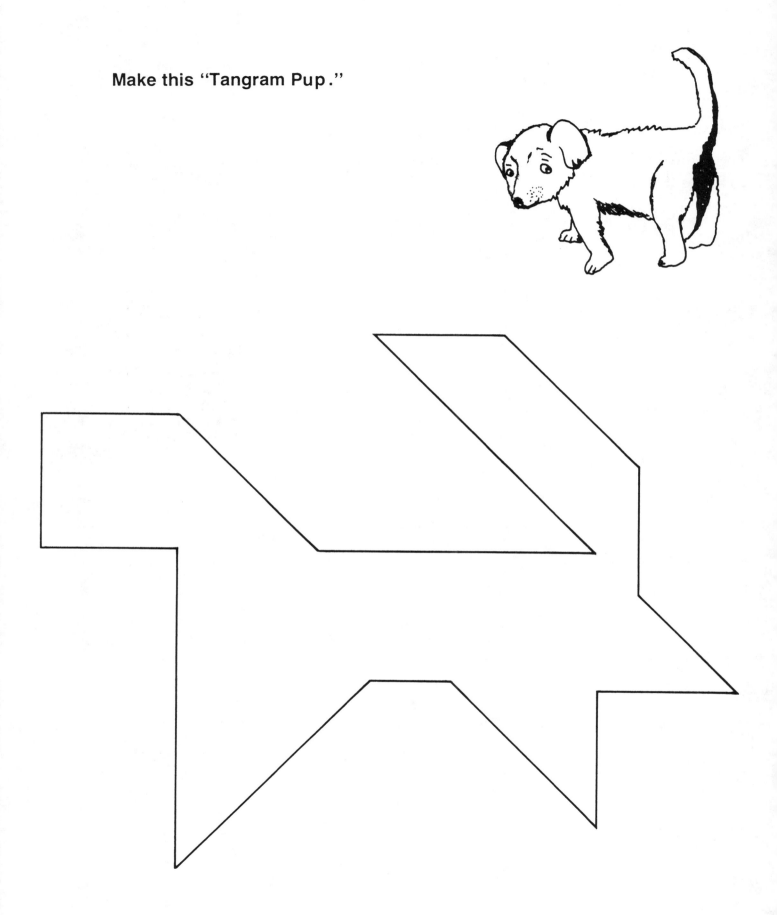

You need all
7 Tangram pieces
to make this Eagle.

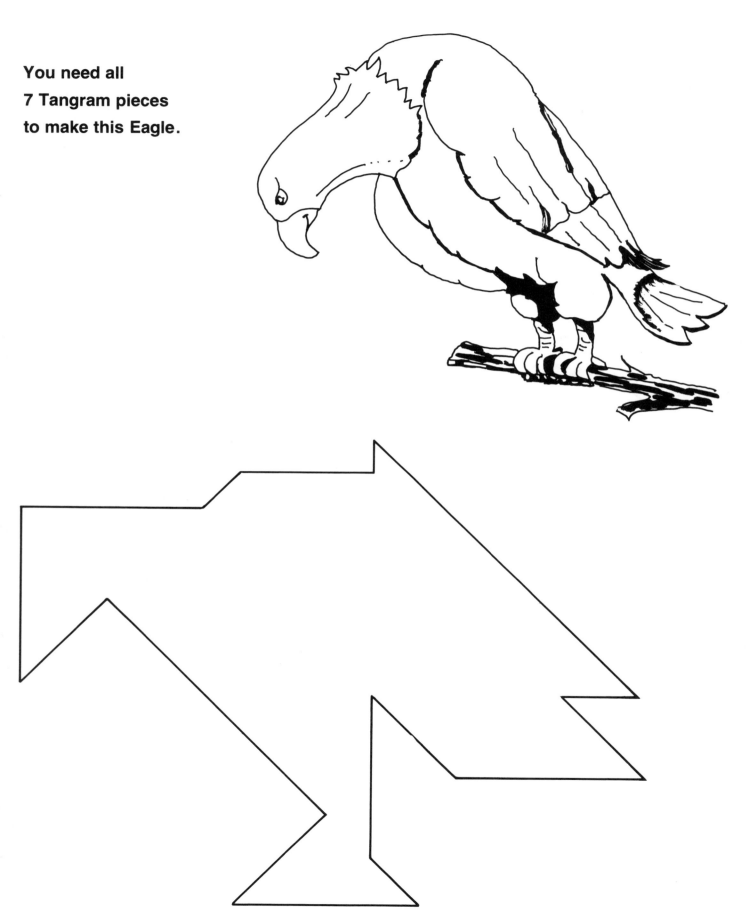

Make a Hare like this using the Tangram pieces.

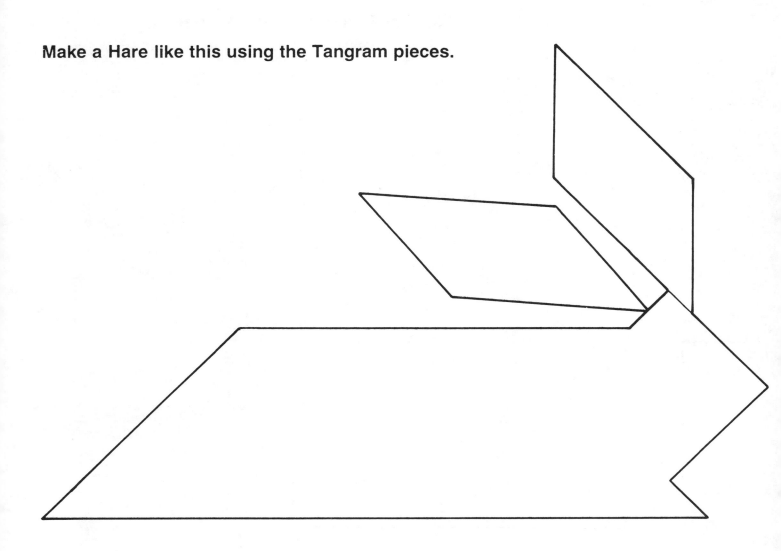

Use the Tangram pieces to make this House.

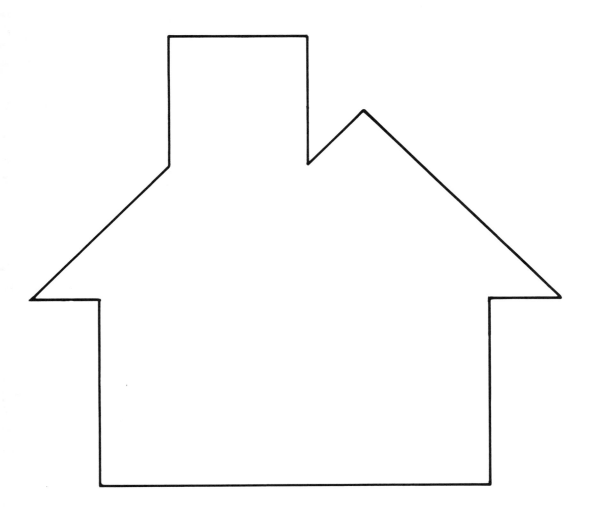

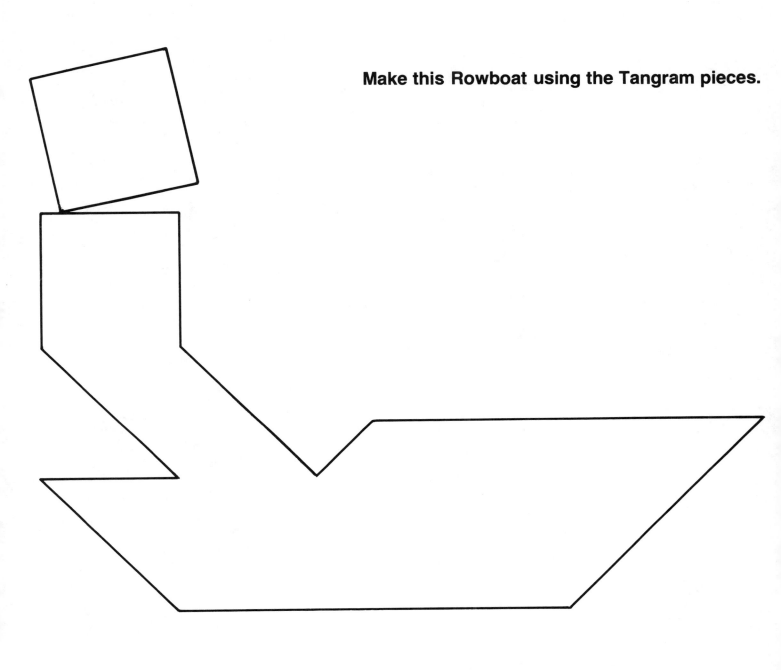

Make this Rowboat using the Tangram pieces.

Use the Tangram pieces to make this Swordfish.

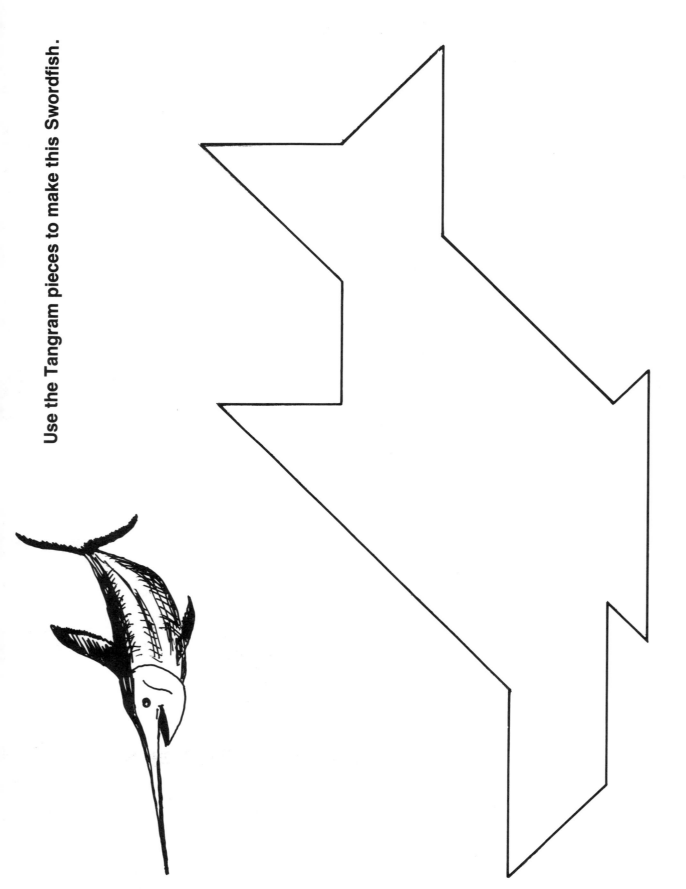

You need all 7 Tangram pieces to make this Parrot.

Make this "Tangram Whale."

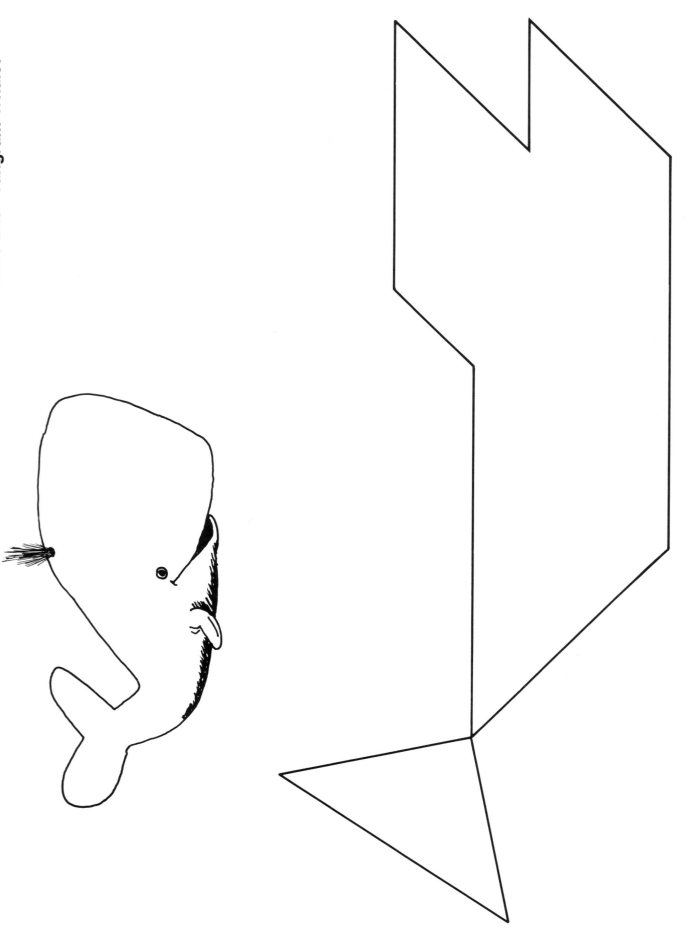

Make this Bull using the Tangram pieces.

You need all 7 Tangram pieces to make this Flagship.

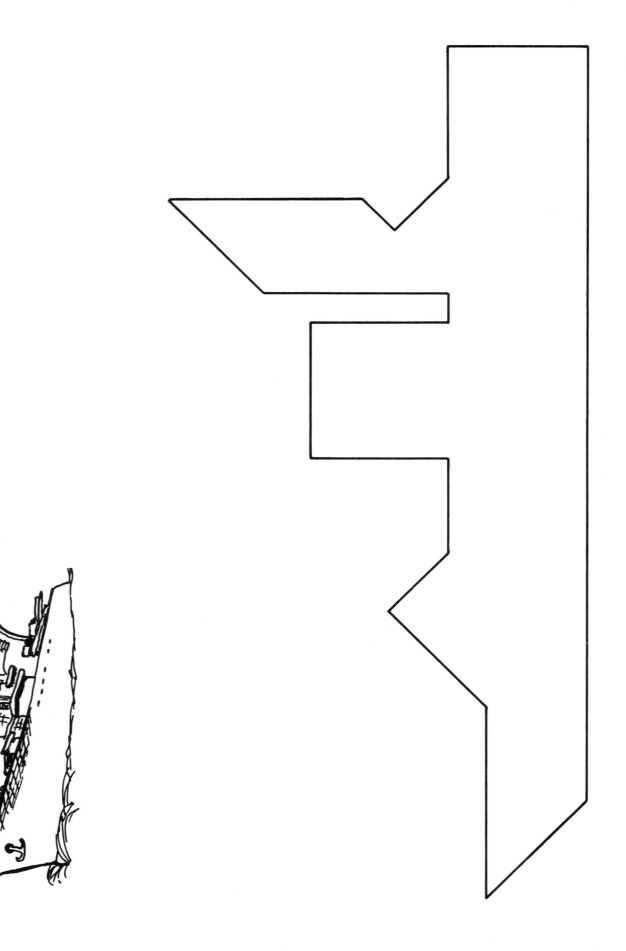

Make this Schooner using the Tangram pieces.

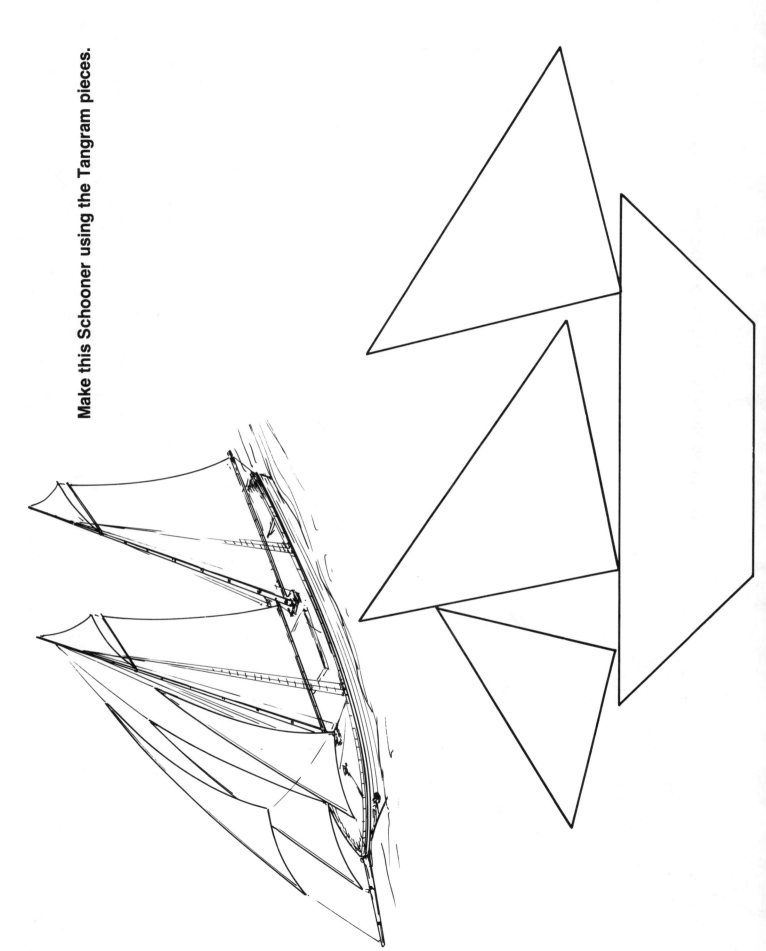

Use the Tangram pieces to make this Fox.

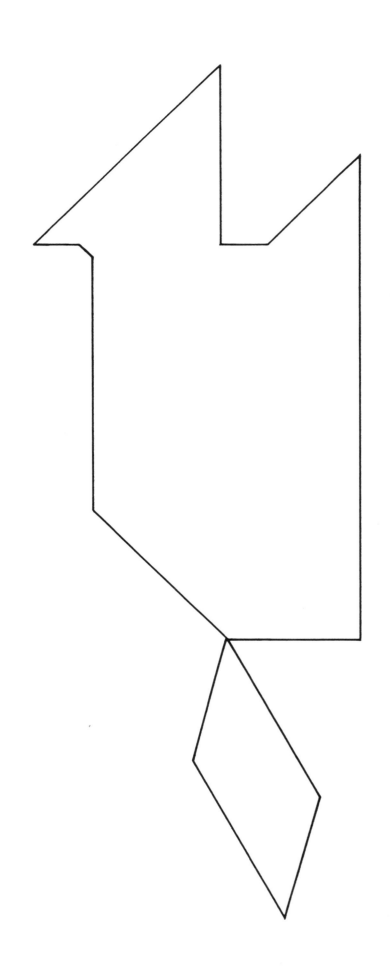

Make a Shark like this using the Tangram pieces.

Can you make this Squirrel using the Tangram pieces?

You need all 7 Tangram pieces to make this Candle.

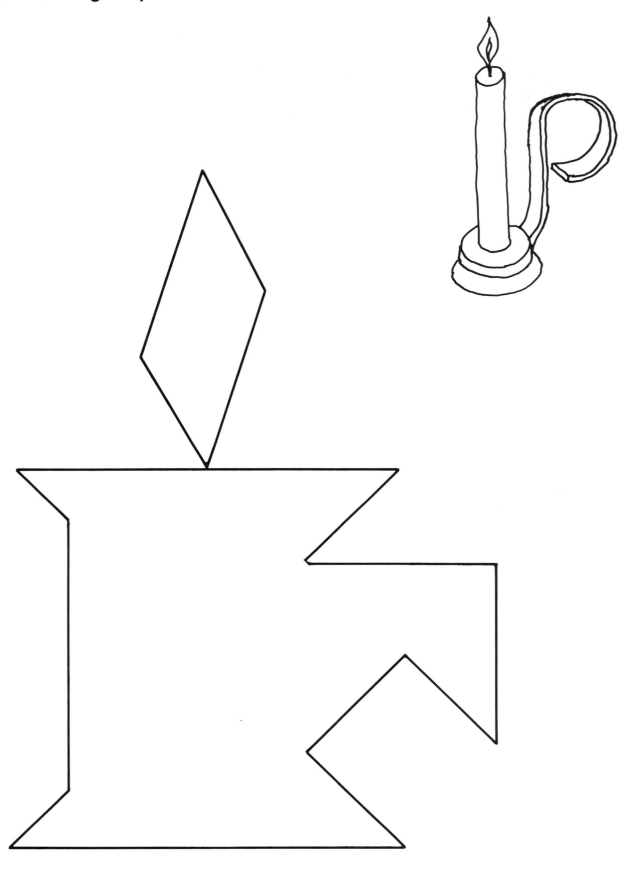

Make this Freighter
using the Tangram pieces.

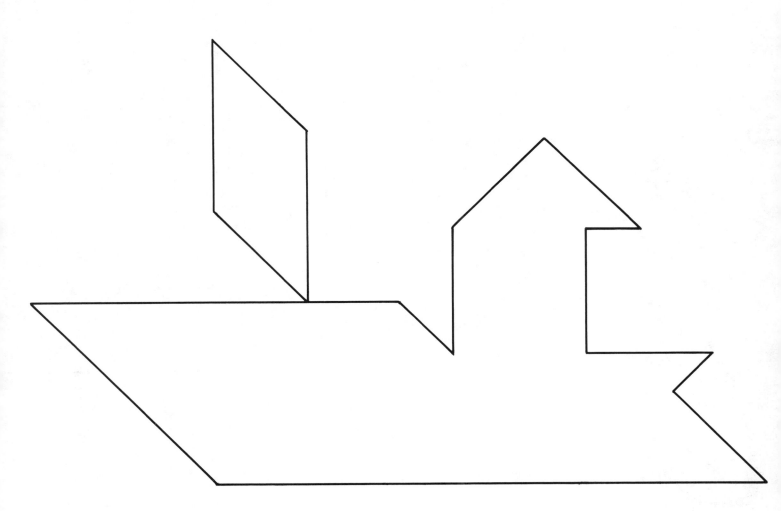

Make a Kangaroo like this using the Tangram pieces.

Make this "Tangram Lobster."

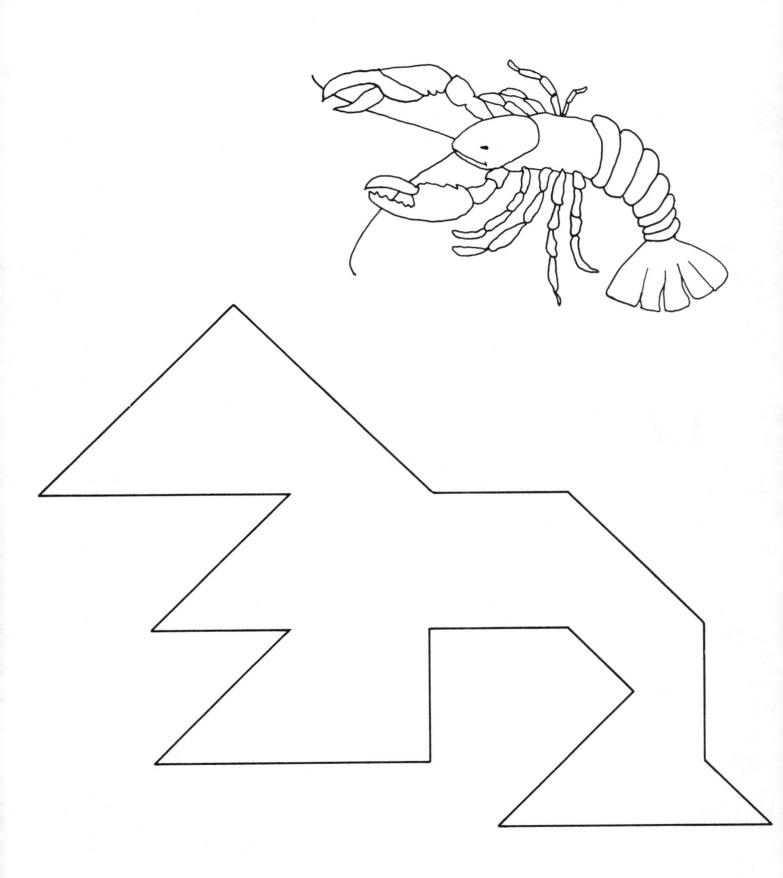

Section Three
Shapes from Shapes

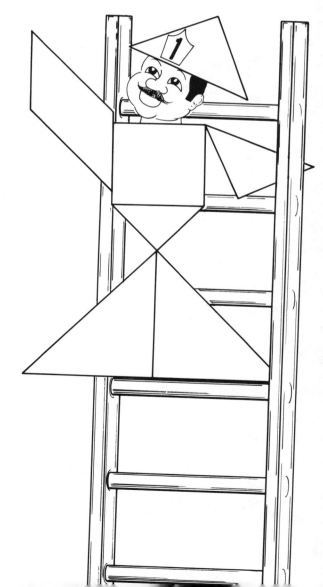

Use these pieces:

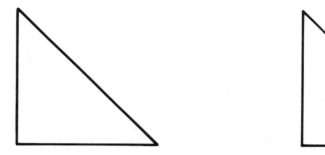

To cover this:

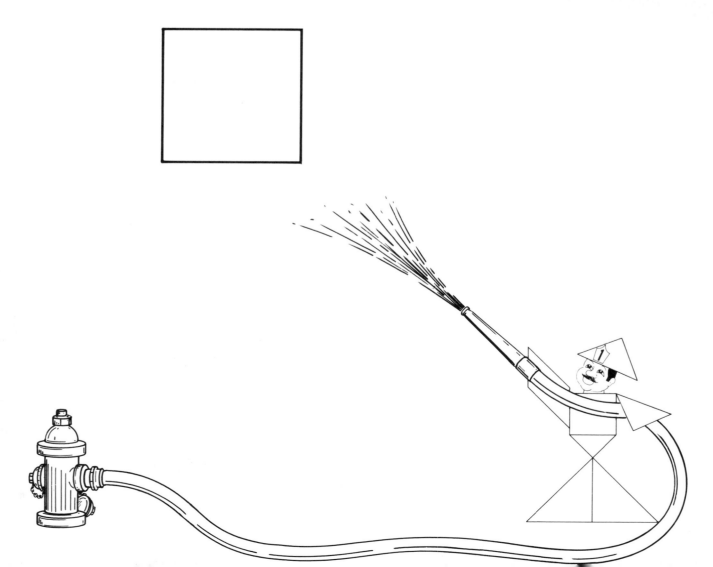

Use these pieces:

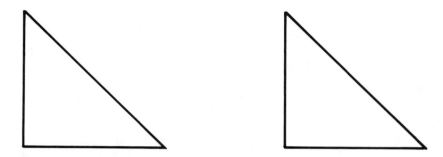

To cover this:

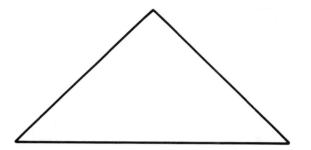

Use them again to cover this:

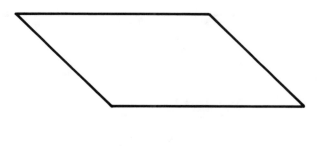

Use these pieces:

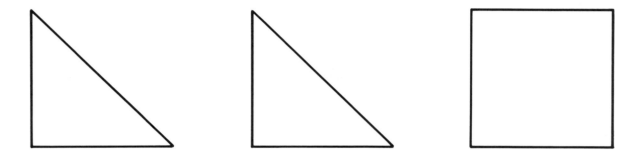

To cover this:

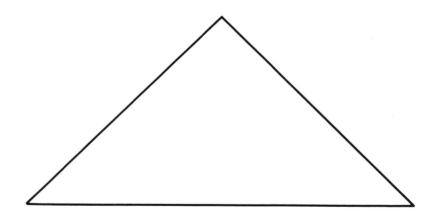

Cover this:

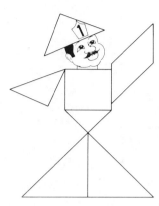 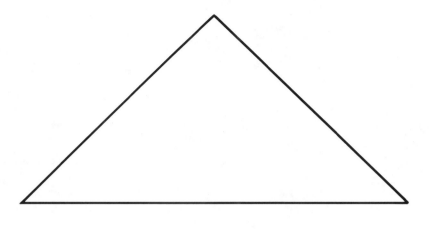

Using these Tangram pieces:

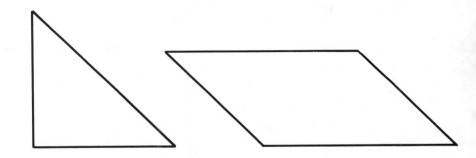

Then use these pieces:

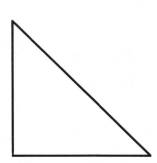 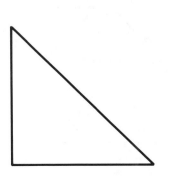 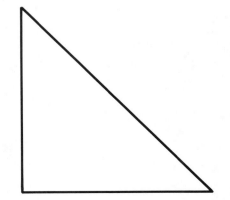

Use these Tangram pieces:

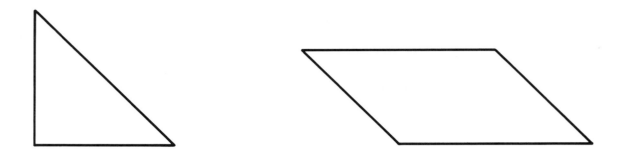

To cover this:

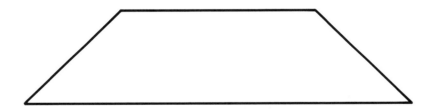

And this:

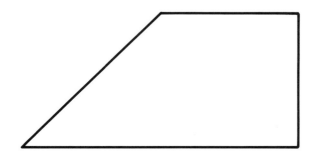

Use these Tangram pieces:

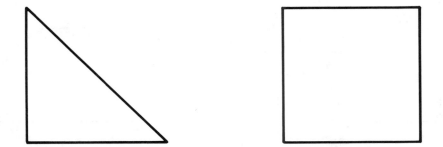

To cover this:

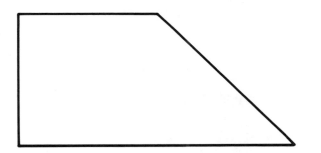

Add one more piece and cover this:

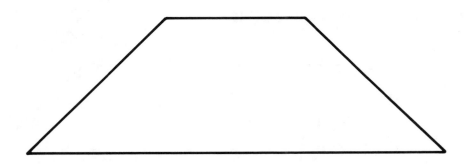

Use these Tangram pieces:

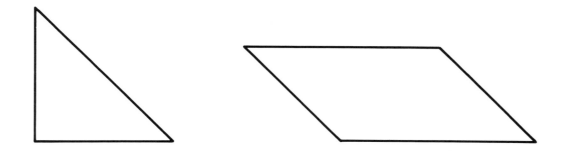

To cover this:

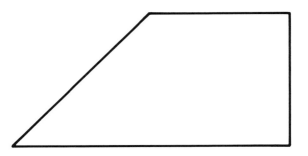

Add one piece and cover this:

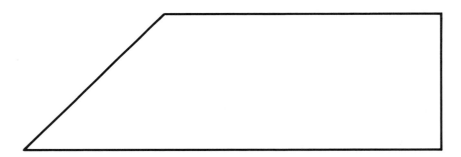

Add one more piece and cover this:

Use two pieces to cover this:

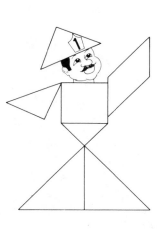

Can you cover it with four pieces?   **Which four pieces? Draw them here.**

Is there another way to cover it with four pieces?

Use two Tangram pieces to cover this:

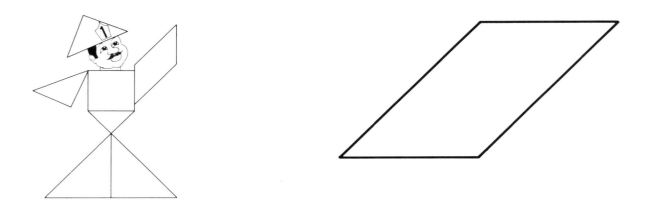

Use two different Tangram pieces to cover this:

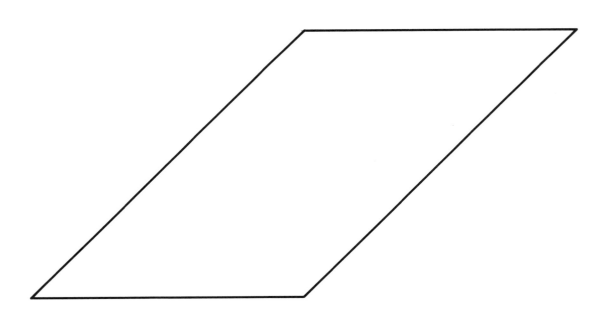

Use three Tangram pieces to cover this:

Use the same three pieces to cover this:

Add one more piece and cover this:

Use five Tangram pieces to cover this:

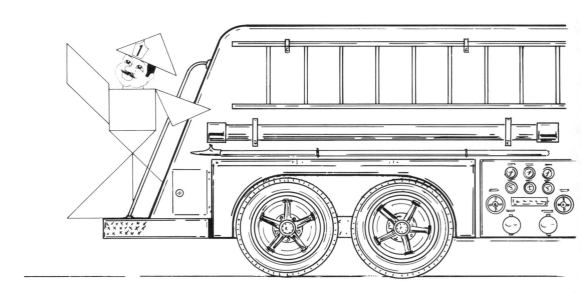

Use six Tangram pieces to cover this:

Cover this using seven Tangram pieces:

Use all seven Tangram pieces to cover this:

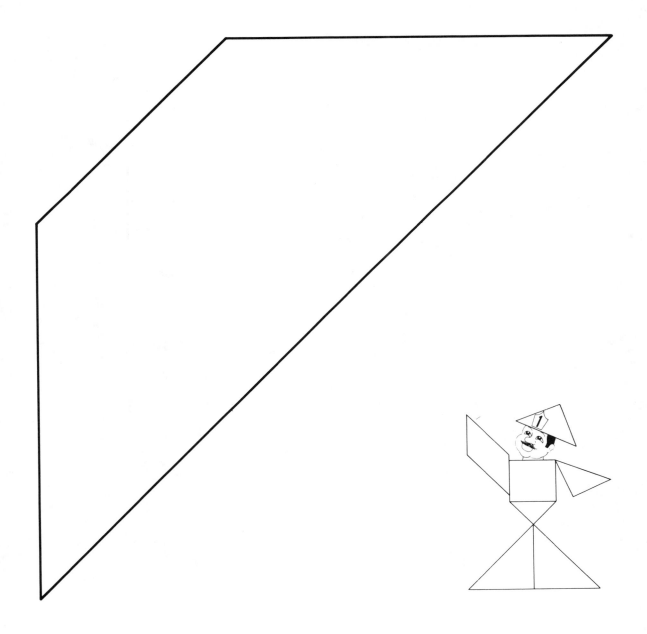

Answers

Section 2 answer key

2-4

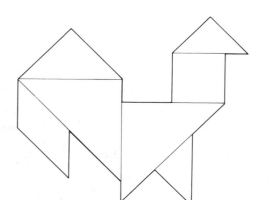

2-5

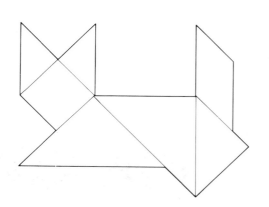

2-6

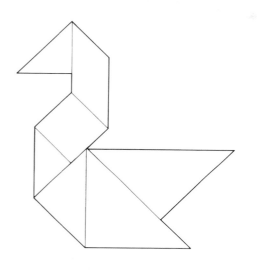

2-7

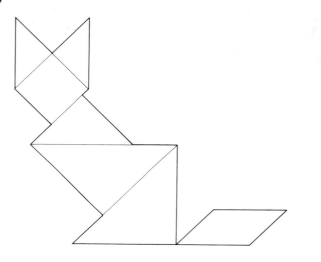

2-8

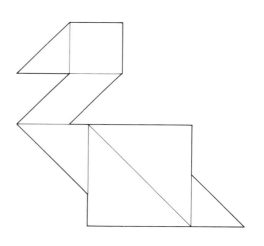

2-9

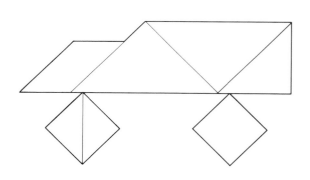

Section 2 answer key continued

2-10

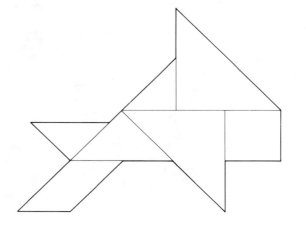

2-11

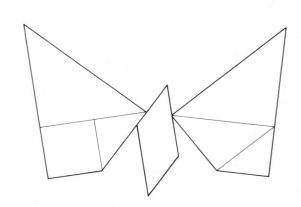

2-12

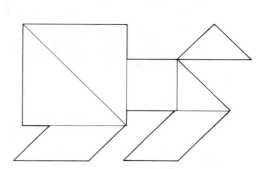

2-13

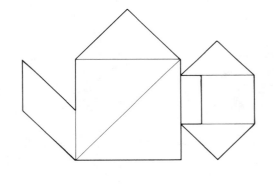

2-14

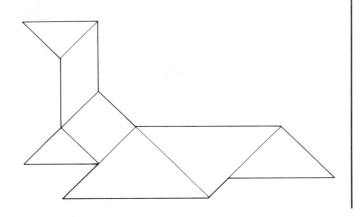

2-15

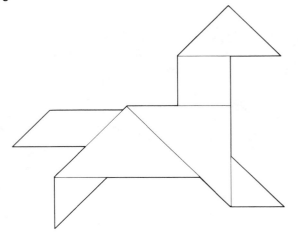

Section 2 answer key continued

2-16

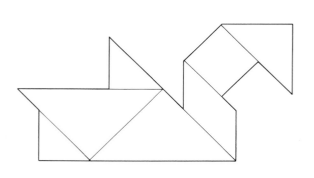

2-17

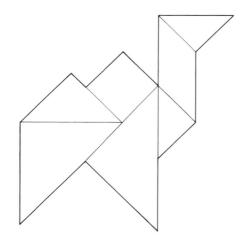

2-18

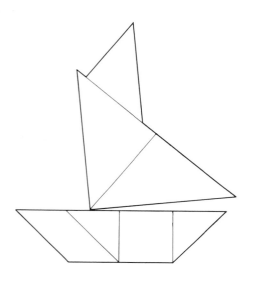

2-19

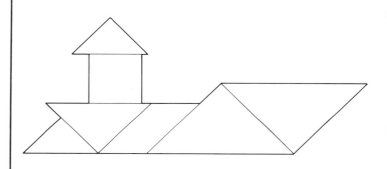

2-20

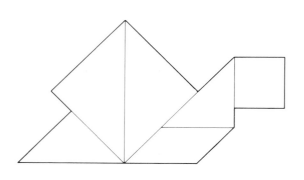

2-21

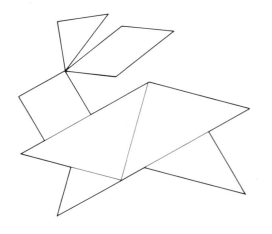

Section 2 answer key continued

2-22

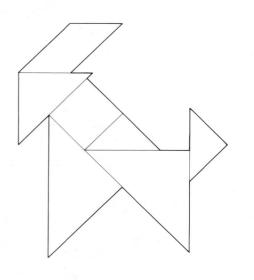

2-23

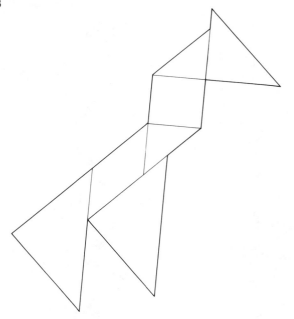

2-24

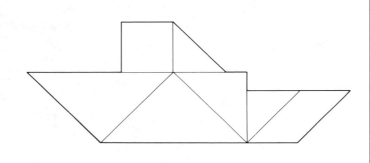

2-25

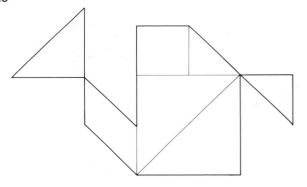

2-26

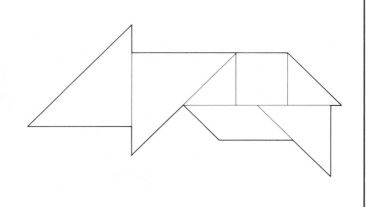

2-27

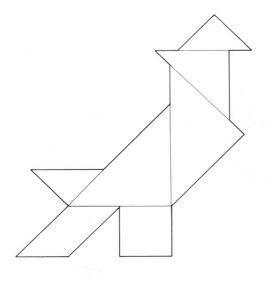

Section 2 answer key continued

2-28

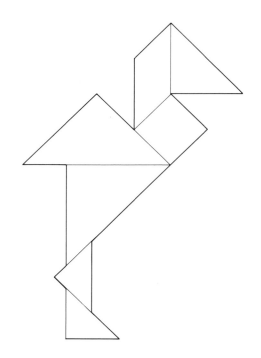

2-29

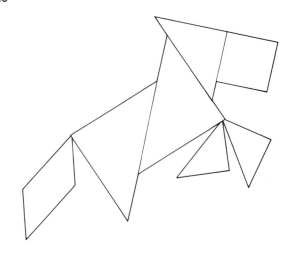

2-30

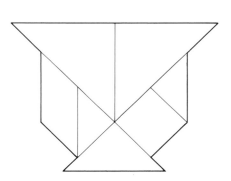

2-31

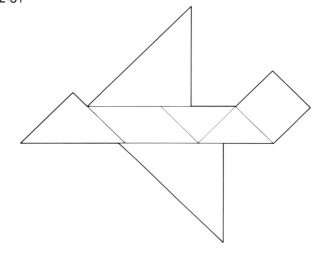

2-32

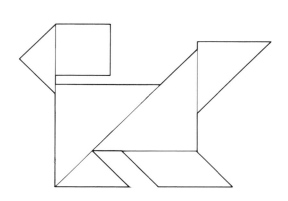

2-33

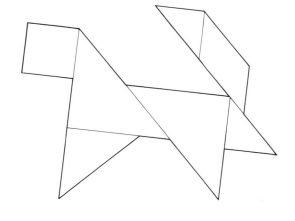

2-34

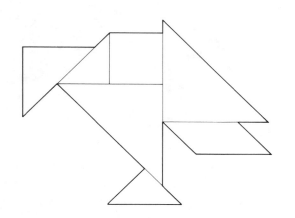

2-35

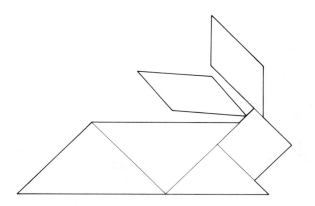

2-36

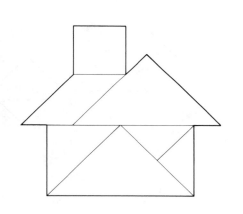

2-37

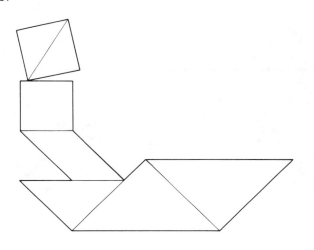

2-38

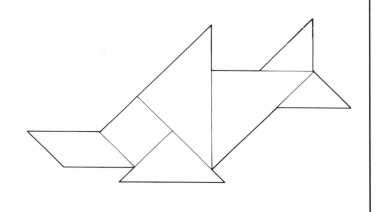

2-39

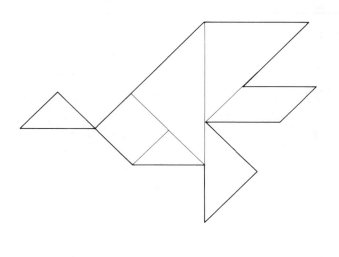

Section 2 answer key continued

2-40

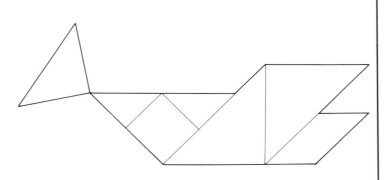

2-41

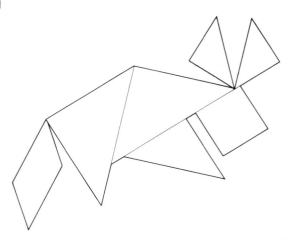

2-42

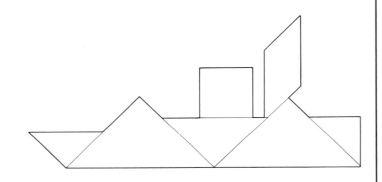

2-43

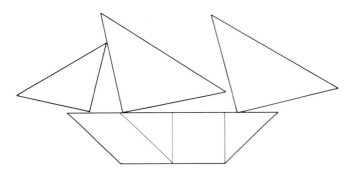

2-44

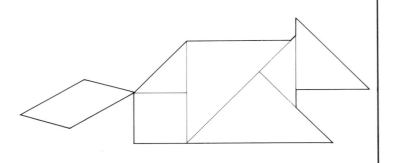

2-45

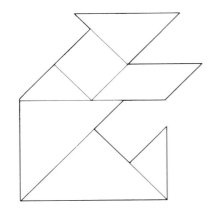

2-46

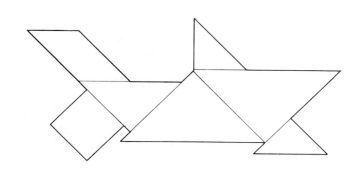

2-47

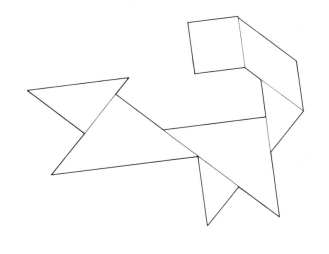

2-48

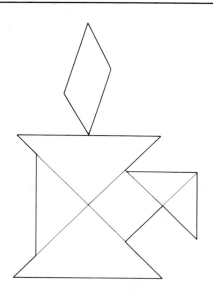

2-49

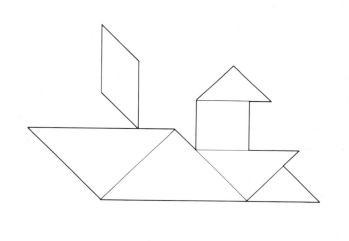

2-50

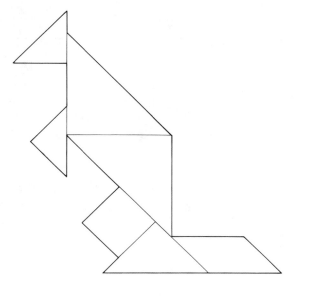

2-51

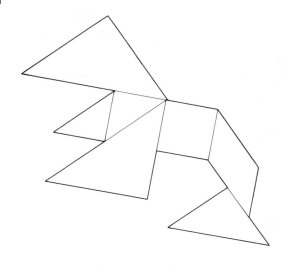

Section 3 answer key

3-2

3-5

3-7

3-3

3-6

3-8

3-4

Section 3 answer key continued

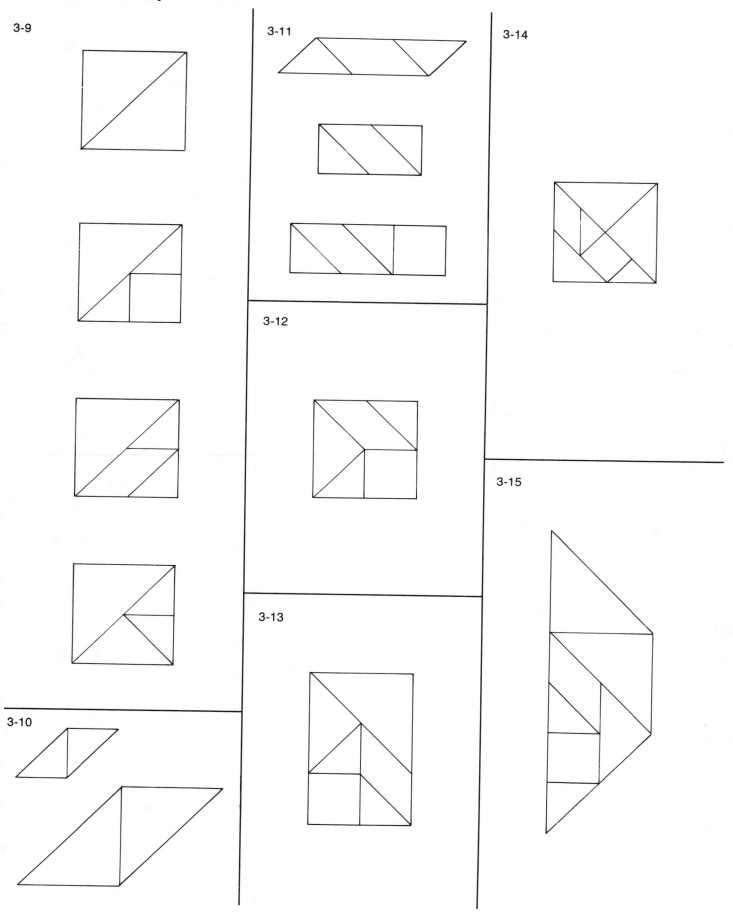

3-9

3-10

3-11

3-12

3-13

3-14

3-15